Table of Contents

1 Introduction to Cooling

Cooling is fundamentally about managing heat transfer, moving thermal energy away from a heat-generating source to maintain a desirable operating temperature. This is crucial for the reliable and extended lifespan of numerous electronic components and systems. Failure to adequately manage heat can lead to performance degradation, malfunction, and even catastrophic failure. Effective cooling designs need to consider various parameters such as the heat generation rate of the component, the ambient temperature, and the desired operating temperature. These considerations dictate the choice of cooling method and components, influencing system effectiveness and cost.

Understanding the relationship between heat and temperature is paramount. Heat is a form of energy, while temperature is a measure of the average kinetic energy of the particles within a substance. Heat transfer always occurs from a region of higher temperature to a region of lower temperature, a principle central to all cooling strategies. This basic thermodynamic principle underscores the core function of any cooling system: to facilitate this heat flow and maintain the operating temperature within acceptable limits. Failing to comprehend this fundamental concept will severely hinder your ability to design effective cooling solutions.

Various cooling methods exist, each with its own strengths and weaknesses. Passive cooling systems rely solely on natural mechanisms like conduction, convection, and radiation to dissipate heat. Examples include heat sinks and strategically designed chassis. Active cooling involves using external power sources, like fans or pumps, to enhance heat removal. This leads to far more efficient cooling than passive systems alone. The selection of an appropriate cooling strategy depends heavily on factors such as the amount of heat generated, the acceptable operating temperature, the available space, and the overall budget constraints. For instance, while air cooling remains a popular choice due to cost-effectiveness, liquid cooling offers superior thermal performance, especially for high-heat density systems.

Air cooling often involves the use of fans and heat sinks. Heat sinks are designed to increase the surface area exposed to the cooling air, thereby enhancing heat transfer through convection. Fans propel air over these increased surfaces, accelerating heat removal. Effective air cooling design involves considerations of fan characteristics such as airflow rate, pressure, and noise levels. Furthermore, careful attention should be paid to the overall system's airflow path, minimizing restrictions and ensuring uniform cooling across all components. Proper design of air ducts and strategically placed vents can significantly improve the efficiency of the system. This is often overlooked in simpler applications but plays a key role in complex cooling arrangements.

Liquid cooling offers a significant advantage over air cooling in terms of heat transfer capacity. Water, because of its high specific heat capacity and thermal conductivity, is frequently employed as a coolant. Liquid cooling systems typically involve a pump to circulate the coolant through the heat source, then a radiator to dissipate heat to the environment. Advanced liquid cooling techniques employ smaller channels for improved heat transfer, enhancing cooling efficiency and enabling the cooling of higher heat flux components. The selection of the coolant and the system design influence the performance and longevity of liquid-cooled systems. Beyond basic water cooling, other liquid cooling methods exist. Refrigerant cooling utilizes refrigerants with highly desirable thermodynamic properties, allowing for efficient heat removal. Refrigerants are typically volatile fluids that change phase as part of the cooling process, further augmenting their cooling potential. These systems, though more complex than water cooling, enable the achievement of very low temperatures and are widely used in industrial and scientific applications. Choosing the right refrigerant depends on factors like desired temperature, environmental impact, safety regulations, and cost considerations. This is crucial for responsible and environmentally sound design.

Phase-change cooling methods exploit the large amount of latent heat associated with phase transitions, such as boiling and condensation. Immersion cooling, where components are submerged in a dielectric liquid, leverages the high heat transfer coefficient during boiling. This offers excellent cooling for high-heat flux components and significantly improves heat dissipation efficiency. Two-phase cooling systems use both liquid and vapor phases to enhance heat transfer, maximizing efficiency while maintaining manageable operating pressures. The choice of the dielectric fluid is critical, requiring consideration of compatibility with the components, safety, and environmental impact. This requires careful selection of cooling medium.

Emerging cooling technologies constantly seek improvements in efficiency and sustainability. Thermoelectric cooling, utilizing the Peltier effect, offers a solid-state alternative to traditional methods. While currently less efficient than conventional techniques for high-power applications, thermoelectric coolers are compact, reliable, and have no moving parts, making them suitable for specific niche applications. The development of novel materials with improved thermal properties promises significant advancements in all cooling techniques. The ongoing research and development in materials science are set to revolutionize heat dissipation, leading to more efficient and sustainable cooling solutions for many applications. Understanding these emerging technologies and their potential is critical for staying at the forefront of the field.

1.1 Cooling Fundamentals

Understanding heat transfer is paramount. It's the bedrock upon which all effective cooling strategies are built. Think of it like this: heat, that unwanted byproduct of your device's operation, wants to escape. It does this through three primary pathways: conduction, convection, and radiation. Conduction involves the direct transfer of heat through a material, like the heat traveling from the processor's die to its heat spreader. Convection relies on fluid movement to carry away heat – air currents rising from a heated surface, or water flowing through a cooling loop. Finally, radiation emits heat as electromagnetic waves, with hotter objects radiating more energy to cooler surroundings. The efficiency of your cooling system hinges on your ability to manage and manipulate these three mechanisms effectively. Understanding their interplay is essential in de-

termining the optimum cooling strategy.
Efficient cooling design doesn't happen by accident; it's a carefully orchestrated dance between components. Choosing the right heat sink, for instance, is crucial. Its surface area directly influences its ability to dissipate heat through convection. A larger surface area allows for more efficient heat transfer to the surrounding air or coolant. Similarly, the material of the heat sink greatly affects its thermal conductivity–how well it transmits heat. Copper and aluminum are popular choices due to their excellent thermal conductivity. Fan selection is equally important. Consider the airflow volume (CFM), static pressure, and noise levels when choosing fans. A higher CFM generally translates to more efficient cooling, but it might also introduce excessive noise. The balance between cooling performance and noise often dictates the best approach for your specific cooling system. Careful consideration of these factors ensures a cohesive and effective system.
Effective thermal management is less about the individual components and more about the system as a whole. Consider the airflow path in air-cooled systems. Proper airflow management ensures that hot air is effectively removed and replaced with cooler air, maximizing heat dissipation. Obstructions within the airflow path, such as poorly placed components or inadequate spacing, hinder this process significantly. This reduction in cooling efficiency can lead to thermal hotspots which significantly reduce component lifespan. In liquid cooling systems, the design of the coolant path is crucial. A well-designed path ensures proper fluid flow, preventing stagnation and maximizing heat transfer from the heat source to the coolant. In both types of systems, the overall system design must be optimized to create a balance between these elements. This

holistic approach is indispensable for achieving optimum cooling performance.

Thermal modeling offers a powerful tool for predicting and evaluating the effectiveness of your design, before you even build a prototype. Software like ANSYS or COMSOL allows you to simulate heat flow within your system, providing valuable insights into temperature distributions, hotspots, and overall system performance. This allows engineers to iterate on their designs quickly and efficiently. These simulations help you identify potential problems early in the process, which are both cost and time-saving measures. Modifying your design based on the model's predictions ensures optimal performance and reduces the need for costly and time-consuming revisions after the construction phase. Through the use of these software, the engineer can be confident in the performance of their system before even undertaking physical construction.

Testing and validation are the final steps in ensuring that your designed system meets the required specifications. This process involves building prototypes and subjecting them to rigorous testing under various conditions. This involves measuring temperatures at different points within the system under typical and extreme operating conditions. The data collected allows you to verify whether your design indeed achieves the targeted temperatures and identifies any unexpected performance issues. Comparing these results with the thermal model data helps refine the model and improve design accuracy in future projects. Careful evaluation is crucial to ensure your design is truly effective and reliable.

Cost analysis is an essential, though often overlooked, aspect of cooling system design. While performance is paramount, you need to balance this with economic considerations. Different cooling methods and components come with varying costs. Weigh the trade-offs between different design choices to optimize performance while staying within budget. This assessment should factor in material costs, manufacturing costs, and long-term operational costs. The most effective cooling solution isn't always the most expensive. Finding the best balance between cooling performance and cost effectiveness is essential for successful product development. Therefore, a careful cost-benefit analysis is vital to ensure the most effective cooling system is chosen.

1.2 Types of Cooling Systems

Passive cooling relies on natural processes to remove heat. Simple designs, like strategically placed vents or heat sinks without active fans, exemplify this approach. While effective for low-power devices or in environments with consistent airflow, passive cooling struggles with higher heat loads, offering limited control over heat dissipation and less efficient for demanding applications. Its simplicity, however, leads to lower initial costs and reduced maintenance, making it a suitable option for certain applications where performance demands are not extreme. The key is understanding the limitations and knowing when this method is appropriate, avoiding situations where it might lead to overheating and system failure.

Active cooling systems directly intervene in the heat removal process. These utilize components like fans, pumps, or compressors to force heat transfer to occur more rapidly and efficiently. Fans, for example, increase convective heat transfer, while pumps circulate coolants like water or refrigerants to absorb heat and transport it away. Active cooling is considerably more efficient than passive cooling, capable of handling much higher heat fluxes and maintaining lower operating temperatures, even under heavy loads. Air cooling, a common type of active cooling, uses fans to blow air over heat sinks, which transfer heat from the device. The effectiveness depends on fan speed and design, air flow management and the surface area of the heat sink. Higher power electronic devices, particularly in tightly confined spaces, rely heavily on active cooling techniques to prevent thermal runaway.

Liquid cooling, another active cooling method, offers superior heat transfer capabilities compared to air cooling. The use of a liquid coolant, typically water or specialized fluids, facilitates efficient heat absorption. This system employs pumps to circulate the coolant, typically through a network of tubes or channels that run close to the heat-generating components. Then, a heat exchanger, such as a radiator, dissipates this heat to the surrounding environment. This methodology allows for a greater heat removal capability than air cooling and more even temperature distribution. High-performance computing, data centers, and industrial applications heavily rely on this cooling approach due to its efficiency and its ability to handle substantial heat loads.

Phase-change cooling employs the significant heat transfer associated with a substance undergoing a phase transition. This process, such as boiling or condensation, involves a large amount of latent heat which enhances heat removal substantially. Immersion cooling, a prominent example, submerges components in a dielectric fluid. When the device heats the fluid, it boils, absorbing substantial heat during the transition to its vapor state. Vapor then condenses, releasing this stored energy. Two-phase cooling systems utilize the dual benefits of liquid and vapor phases for optimal efficiency. Phase-change cooling demonstrates significant advantages for high-heat-flux applications, where traditional air or liquid cooling systems may prove insufficient; however, these systems require meticulous design and component selection. Understanding the thermodynamic properties of the working fluid is essential for optimal performance.

Thermoelectric cooling presents a different approach. It utilizes the Peltier effect, where an electric current running through a junction of two dissimilar conductors induces a temperature difference. This enables the targeted cooling of specific areas by precisely managing the electric current through the thermoelectric module. While offering excellent temperature control and localized cooling capabilities, thermoelectric coolers suffer from relatively low efficiency compared to more traditional cooling mechanisms. Their efficiency is generally lower than conventional cooling techniques, making them a preferred choice in specific niche applications. Further advancements in thermoelectric materials and enhanced cooling designs might broaden their applicability in the future. Each system type presents a unique trade-off between cost, efficiency, complexity and suitability for specific applications, calling for a careful consideration of the overall system needs.

2 Heat Transfer Mechanisms

Conduction, the simplest heat transfer mechanism, relies on direct contact. Imagine holding a hot cup of coffee; the heat transfers directly from the cup to your hand. This transfer is governed by Fourier's Law, a cornerstone equation in thermal engineering: the rate of heat transfer is proportional to the temperature gradient and the thermal conductivity of the material. High thermal conductivity materials, like copper or diamond, facilitate rapid heat transfer, making them ideal for heat sinks. Conversely, insulators, such as air or specialized polymers, possess low thermal conductivity and are used to minimize unwanted heat flow. Understanding conductive heat transfer is crucial for choosing materials in your cooling system; the choice of materials directly impacts the effectiveness of your cooling strategy. Improper selection can lead to hotspots, thermal runaway, and ultimately, device failure. Effective management of conduction requires careful consideration of material properties and efficient pathways for heat dissipation, often relying on high-surface-area designs to maximize contact and minimize thermal resistance.

Convection, significantly more complex than conduction, involves heat transfer through fluid motion. This process differs from conduction in that heat is transported by the bulk movement of a fluid, either a liquid or a gas. Consider a radiator; as the coolant flows through it, it absorbs heat from the components and carries it away, dispersing the heat into the surrounding environment. This movement is driven by temperature differences, creating density gradients that induce buoyancy-driven convection, or through forced convection using fans or pumps. The efficiency of convective heat transfer hinges on the fluid's properties (viscosity, thermal conductivity, specific heat), flow rate, and the geometry of the heat transfer surface. Increasing the surface area, for example through fins, improves heat transfer, as does using a fluid with high thermal conductivity and a low viscosity. Efficient convective cooling systems are optimized to maximize both surface area and fluid flow to provide robust heat dissipation. This is precisely why you need to understand the flow characteristics of your system and the impact of pressure drops and turbulence.

Radiation, unlike conduction and convection, doesn't require a medium. It's the electromagnetic wave emission of thermal energy from a surface. The rate of radiative heat transfer depends heavily on the surface's temperature and emissivity—a material's ability to emit thermal radiation. A blackbody, a theoretical perfect emitter and absorber, exhibits an emissivity of 1. Real-world objects have emissivities less than 1. The Stefan-Boltzmann Law quantifies this relationship, stating that radiative heat flux is proportional to the fourth power of the absolute temperature. This means that even small temperature increases result in significant increases in radiated heat. This is why radiative heat transfer becomes increasingly significant at higher temperatures. Effective cooling designs often incorporate strategies to minimize radiative heat gain, such as using reflective surfaces or isolating components, especially in environments with high ambient temperatures or significant radiant heat sources. Consider the design of a server room, where heat from numerous servers is a significant concern. Designing for effective heat dissipation involves mitigating radiant heat transfer from the environment and the components themselves.

In designing efficient cooling systems, a holistic approach considering all three heat transfer mechanisms is essential. Imagine designing a laptop cooling system: the heat generated by the processor is initially transferred via conduction to the heat spreader. From there, convection, facilitated by a fan, carries the heat away, and radiative heat loss plays a secondary role, with minimal contribution when compared to the other two modes. However, understanding the interplay of these mechanisms is vital. A system might excel in convective cooling, but poor conduction from the heat source to the heat sink can significantly reduce overall effectiveness. Thus, a comprehensive cooling strategy considers every pathway of heat transfer, selecting materials and geometries that optimize each mode to create an effective and efficient system. Poor understanding of this interaction is one of the most prevalent design flaws, leading to overheating and malfunction.

Therefore, mastering heat transfer fundamentals is not merely an academic exercise; it is a cornerstone skill for designing effective and reliable cooling systems. A deep understanding of conduction, convection, and radiation enables engineers to make informed decisions about material selection, component design, and overall system architecture. Ignoring any of these three modes may result in suboptimal cooling and, potentially, catastrophic system failure. The proper utilization of these principles in conjunction with an awareness of fluid dynamics and thermodynamics empowers engineers to develop cooling solutions suitable for diverse applications, from microelectronics to large-scale industrial processes, ensuring operational reliability and extended device lifespan.

2.1 Conduction

Conduction relies on the direct transfer of thermal energy through a material. Imagine holding a metal spoon in a hot cup of tea; the heat rapidly travels up the spoon's handle, making it too hot to touch. This is conduction in action. The speed at which this heat travels depends heavily on the material's thermal conductivity—a property representing its ability to conduct heat. Materials like copper and aluminum boast high thermal conductivity, making them excellent choices for heat sinks and other cooling components, while materials like air possess much lower conductivity, making them poor conductors. Understanding this fundamental difference is crucial for effective cooling system design; selecting materials with high thermal conductivity allows for efficient heat removal from a heat-generating source. The efficiency of conduction is heavily dependent on various factors, including temperature gradient and surface area. Effective management of conductive heat transfer often involves optimizing the thermal pathways. This includes selecting materials with high thermal conductivity, creating efficient interfaces between components (minimizing contact resistance), and maximizing the contact area between the heat source and the heat sink. For instance, the application of thermal interface materials (TIMs) like thermal grease or pads between a CPU and its heat sink dramatically reduces contact resistance, greatly improving heat dissipation. Similarly, the design of heat sinks, characterized by their extended surface area and often intricate fin structures, significantly enhances heat transfer by conduction from the device to the surrounding environment. A well-designed heat sink acts as an intermediary, increasing the surface area available for heat exchange and improving the overall effectiveness of the cooling system.

Considering the material properties and geometrical aspects is not sufficient, however. Conduction also hinges on the temperature difference between the heat source and the heat sink; a larger temperature gradient means more rapid heat flow. This underscores the importance of maintaining a low operating temperature for the device and ensuring efficient heat transfer to the coolant (air or liquid). Think of it as a water flowing downhill; the steeper the slope, the faster the water flows. Similarly, a steeper temperature gradient leads to faster heat flow through conduction, thereby enhancing the effectiveness of the cooling mechanism. This principle guides the selection of efficient cooling methods.

To optimize a design, often thermal modeling becomes necessary. Sophisticated software tools, employing techniques like finite element analysis (FEA), allow engineers to accurately simulate heat flow and predict the temperature distribution within a component or system. This predictive capability provides invaluable insights into the effectiveness of the design, potentially revealing "hot spots" or areas of insufficient cooling. These models allow for iterative design improvements and optimization even before physical prototyping, considerably reducing development time and costs. Using such simulations, you can make informed choices about material selection, geometry optimization, and the size and placement of heat sinks to maximize the efficiency of conductive heat transfer. This process leads to a more robust and reliable design, particularly crucial for electronics where excessive heat can cause malfunctions or damage. Effective conduction, therefore, is not merely a matter of choosing the right materials but also involves careful consideration of the design parameters, including contact resistance, surface area, and temperature gradients. Analyzing these factors requires a blend of theoretical knowledge and practical considerations, resulting in a design tailored for the specific demands of the application. Advanced analytical techniques and simulation tools allow engineers to fine-tune their designs and push the boundaries of cooling efficiency. By understanding and optimizing conduction, we can achieve more efficient and reliable cooling systems, ensuring the optimal functionality and longevity of heat-generating devices. The impact reaches various applications; from high-performance computing to everyday electronics, optimized conduction is key to performance and reliability.

2.2 Convection

Convection relies on fluid movement to transfer heat. Imagine a pot of boiling water; the hotter water at the bottom rises, displacing cooler water above it, creating a continuous cycle. This cyclical motion, driven by density differences caused by temperature variations, is the essence of natural convection. Understanding this fundamental principle is crucial for designing effective cooling systems, especially in scenarios where the heat source generates substantial thermal energy. Efficient cooling necessitates careful consideration of the fluid's properties, such as viscosity and thermal conductivity, as these factors directly influence the rate of heat transfer. The design of the cooling system must facilitate and enhance this natural convection process, for example, by ensuring optimal flow paths and minimizing obstructions. Failure to do so can result in stagnant fluid zones and reduced cooling efficiency.

Forced convection, a more active method, uses external means to enhance heat transfer. This commonly involves fans or pumps that actively circulate the cooling fluid, significantly accelerating the heat dissipation process. Forced convection systems are prevalent in electronic devices, data centers, and industrial machinery. The selection of a suitable fluid is paramount. Water, due to its high specific heat capacity, is a popular choice, but its inherent limitations might necessitate the use of specialized fluids like refrigerants or engineered coolants, especially when dealing with high-power applications that generate extreme temperatures. The engineering aspects are significant, encompassing the design of heat exchangers, fan characteristics, and even the placement of components to maximize the impact of the forced flow. Poorly designed systems might lead to uneven temperature distribution, localized overheating, and ultimately, system failure.

Beyond the choice of fluid and the method of circulation, the geometry of the cooling system plays a critical role in convection's effectiveness. Consider the design of a heat sink; its fin arrangement directly impacts surface area available for heat transfer. A larger surface area allows for more efficient heat dissipation through convection, whether natural or forced. Similarly, the spacing between fins influences the fluid's flow pattern, impacting its effectiveness. Too close and the fluid may struggle to move freely, reducing convective heat transfer. Too far apart and the heat transfer area is reduced. The optimal spacing depends on the specific fluid and the convective heat transfer regime. Precise design requires rigorous thermal modeling and simulations.

The interplay between natural and forced convection is often subtle but crucial. In many systems, both mechanisms contribute simultaneously. Natural convection might establish a base level of cooling, while forced convection enhances it significantly. Designing such systems requires a holistic understanding of how these two modes interact. The balance between natural and forced convection is not merely a matter of adding a fan; it involves understanding how the fan's flow interacts with the natural convection currents. Careful design might leverage these combined effects to optimize overall cooling performance while minimizing energy consumption and system complexity. This complex interaction necessitates a deeper understanding and requires sophisticated computational fluid dynamics (CFD) analysis to accurately predict the temperature distribution and validate the system's efficacy.

Furthermore, the boundary conditions in a convection system dramatically influence its performance. These conditions define the thermal environment surrounding the system, influencing how heat is exchanged with the surroundings. Things like ambient air temperature, airflow patterns external to the system, and the thermal properties of surrounding materials all affect the overall heat transfer. Understanding and incorporating these external factors into the design is essential to ensure that the cooling system performs consistently and reliably in its intended operating environment. Ignoring these environmental factors can lead to unexpected results, rendering even carefully designed systems ineffective. A well-engineered cooling system should account for the full thermal environment, predicting and mitigating the influence of these external factors.

2.3 Radiation

Radiation, unlike conduction and convection, doesn't require a medium for heat transfer. Energy travels as electromagnetic waves, directly from a hot surface to a colder one. This makes it a particularly relevant cooling mechanism in situations where a vacuum exists, or where minimizing contact is crucial–think of spacecraft thermal management or high-vacuum applications. Understanding its principles is vital for optimizing cooling system performance in such scenarios, and even in more commonplace situations where radiative heat transfer can contribute significantly to overall heat dissipation. We need to consider its intricacies to make informed choices during the design process.

The intensity of radiative heat transfer depends heavily on the surface properties of the involved objects. Emissivity, a crucial parameter, represents a material's ability to emit thermal radiation. A higher emissivity means more efficient radiation heat transfer. Conversely, reflectivity–the ability to reflect incident radiation–plays a significant role in minimizing heat gain, a critical design consideration, especially when designing for environments with significant external heat sources. Proper material selection, therefore, is not merely about thermal conductivity or specific heat capacity but also encompasses radiative properties. This adds another layer of complexity, particularly for systems exposed to varying environmental conditions, requiring detailed analyses to predict and manage radiative heat fluxes.

Consider the design of a high-power laser diode. Efficient cooling is essential for preventing overheating and maintaining performance. While convection and conduction help, radiation plays a noteworthy role, particularly in removing heat from the laser diode's active region. A heat sink with a high emissivity coating will drastically improve the efficiency of radiative cooling, and its placement and design should be considered carefully. Using a material with high thermal conductivity to quickly transport heat from the diode to the heat sink's surface is beneficial. Strategically positioning the laser within an enclosure with controlled reflectivity can further augment the system's overall cooling performance. This careful attention to detail highlights the importance of understanding and harnessing radiative heat transfer for optimizing system longevity and performance.

Now, let's discuss the Stefan-Boltzmann law. It dictates the power radiated per unit area by a blackbody as a function of its absolute temperature. While no real-world material is a perfect blackbody, this law provides a valuable framework for understanding radiative heat transfer, especially in simplified models. The law highlights the strong dependence of radiated power on temperature; a small increase in temperature leads to a significant rise in radiated power. In designing cooling systems, this emphasizes the critical importance of keeping the system's temperature as low as possible. The law also forms the basis for several more complex calculations, useful in refining theoretical heat transfer models for more complex scenarios and components. These refinements are crucial to accurate predictions of real-world cooling performance, ensuring the cooling system achieves the desired outcomes.

Beyond the Stefan-Boltzmann law, the view factor plays a critical role. It represents the fraction of radiation leaving one surface that strikes another, and it considers the geometry and relative orientation of the surfaces involved in the radiative exchange. Accurately calculating view factors is essential for precise thermal modeling. Software tools and computational techniques are frequently used to evaluate these factors, particularly in complex geometries where analytical solutions become intractable. This means advanced computational methods are essential for designing efficient cooling systems. Their utilization is vital for achieving the level of precision needed for optimal system design in the face of complex geometric relationships between components.

Let's illustrate this with an example. Imagine a server rack densely packed with components. The view factors between the various components heavily influence the radiative heat transfer between them. A proper thermal model must carefully consider these factors to accurately predict temperatures and optimize the placement of components. The inclusion of these factors in the modelling process enhances accuracy and precision, allowing for fine-tuning of designs and promoting greater efficiency. This is crucial for minimizing overall operating temperatures and improving the longevity and reliability of the server's components. Neglecting these factors can lead to erroneous predictions and suboptimal system design, resulting in increased risk of failure.

Finally, consider the impact of surface coatings. Selective surfaces, engineered to have tailored emissivity and reflectivity in specific wavelength ranges, can significantly influence radiative heat transfer. Such coatings can enhance radiative cooling by maximizing emission at specific wavelengths, while minimizing absorption at others. This targeted approach allows for the creation of systems that are more efficient at dissipating heat, and it can be particularly beneficial in applications where specific environmental conditions prevail. The selection of the appropriate coating depends heavily on factors such as the operating temperature, the surrounding environment, and the system's specific requirements. Thorough experimentation and analysis are typically necessary to determine the most effective coating for a given application. These tailored surfaces can improve cooling efficiency significantly, providing a more effective solution for managing heat dissipation and prolonging the lifespan of the system.

3 System Design Principles

Component selection forms the bedrock of any successful cooling system. Choosing the right heat sink, for instance, depends heavily on the heat generated by the device, the available space, and the desired operating temperature. Oversizing a heat sink might seem beneficial, but it unnecessarily increases costs and system bulk. Conversely, undersizing it leads to insufficient cooling, potentially damaging sensitive components. Careful consideration of material properties, fin density, and surface area is crucial. Similarly, fan selection requires evaluating airflow rate, static pressure, and noise levels against the heat sink's thermal resistance and the ambient environment. A high-performance fan in a poorly designed airflow pathway offers minimal advantage; equally, a low-performance fan on a highly efficient heat sink can cause overheating. This intricate interplay emphasizes the importance of considering each component's contribution to the overall system performance. The process is iterative; initial selections often need adjustments based on subsequent modeling and testing.

Effective thermal modeling is paramount. It predicts system performance before physical prototyping, saving time and resources. Finite element analysis (FEA) is a powerful tool simulating heat transfer within complex geometries. By defining boundary conditions such as ambient temperature, heat generation rates, and material properties, FEA software can accurately estimate temperature distributions throughout the system. This process isn't simply about plugging numbers into a software package; it requires expertise in defining accurate boundary conditions. Incorrect estimations can lead to unreliable predictions. The model's mesh density also impacts accuracy; a finer mesh improves precision but dramatically increases computational cost and time. A balance must be found between accuracy and efficiency. The results from the thermal model are crucial for informed decisions regarding component placement, airflow pathways, and material choices. It's an essential step in optimizing system performance and preventing potential failures.

Optimization strategies aren't just about improving performance—they're about finding the optimal balance between performance, cost, size, and other relevant factors. Consider the objective function to quantify desired performance, perhaps minimizing the maximum temperature, or maximizing thermal efficiency. Then, carefully select variables you can control; these might include heat sink dimensions, fan speed, coolant flow rate, or the use of phase-change materials. Various optimization algorithms exist for this process, ranging from gradient-based methods, suitable for smoother objective functions, to evolutionary algorithms which better handle complex, discontinuous landscapes. Software tools can streamline this; however, human expertise remains critical for formulating the problem, interpreting results, and making informed choices. An unoptimized system might achieve adequate cooling but at the expense of increased size, weight, or power consumption; optimization prevents such inefficiencies.

Thorough testing and validation are indispensable steps; a well-designed cooling system on paper might still fail in practice. Testing should simulate real-world operating conditions; this includes variations in ambient temperature, device workload, and potentially unexpected environmental conditions. Data acquisition systems, often employing temperature sensors and flow meters, are critical in collecting reliable data. Comparative analysis against the model's predictions provides valuable insights. Discrepancies might highlight areas needing redesign or refinement. Experimental validation builds confidence in the system's performance and reliability. It's a crucial step in demonstrating the system's ability to withstand operational stresses and maintain its designed effectiveness under various scenarios.

Cost analysis is often overlooked, but it's crucial. It isn't simply about adding up the costs of individual components; it's about evaluating the cost-effectiveness of different design options. This requires considering various factors such as material costs, manufacturing costs, energy consumption over the system's lifetime, and potential maintenance expenses. Life-cycle cost analysis provides a comprehensive perspective. Sometimes, a slightly more expensive component might lead to considerable savings in the long run through increased reliability or reduced energy consumption. This detailed analysis helps make rational decisions. Optimizing for cost without compromising functionality is an essential aspect of producing a commercially viable and sustainable cooling solution. The goal is not just to design a functional system but also a financially responsible one that balances performance, durability, and cost-effectiveness.

3.1 Component Selection

Choosing the right components is paramount; it directly impacts the system's efficiency, reliability, and overall cost. A poorly selected component can significantly hinder performance, even rendering the entire system ineffective. Therefore, meticulous consideration is essential from the outset.

The selection process isn't simply choosing the most powerful or expensive option. Instead, it requires a balanced approach that considers various factors, including the thermal characteristics of the device being cooled, the ambient environment, and the desired cooling performance. For example, a high-power CPU might necessitate a larger, more robust heatsink and a high-flow fan, while a low-power device may only need a simple heatsink and passive cooling.

Understanding the heat dissipation requirements of your device is foundational. This involves accurately determining the power dissipation, which often requires consulting datasheets, performing thermal simulations, or even experimental measurements. The heat flux, or heat flow rate per unit area, is a crucial parameter that guides component sizing and selection. For instance, a device with a high heat flux demands components with a high thermal conductance to ensure effective heat transfer.

Heat sinks are the workhorses of many cooling systems. Their primary function is to increase the surface area available for heat dissipation, enhancing convection and radiation. Material selection plays a crucial role; aluminum is common due to its high thermal conductivity and cost-effectiveness, while copper offers even higher conductivity for situations demanding superior heat transfer. The fin design significantly affects performance; larger surface areas generally translate to better cooling, but design must consider air resistance.

Fans play a vital role in many active cooling systems by creating airflow to enhance convective heat transfer. Key parameters include airflow rate (CFM), static pressure, and noise levels. Selecting a fan with adequate airflow to match the heatsink's design is crucial. Oversized fans might be quieter but less efficient, while undersized ones may not provide sufficient cooling. The fan's location and orientation also influence its effectiveness.

For liquid cooling systems, pump selection is critical. The pump must provide sufficient flow rate to circulate the coolant efficiently through the system, overcoming pressure drops in the tubing and radiator. The pump's reliability is equally important. Factors like pressure head, flow rate, and noise levels are all important considerations. The pump's material compatibility with the coolant is also vital to prevent corrosion and leakage. Furthermore, the pump's power consumption needs to be factored into the overall system energy efficiency calculations.

Tubing, often overlooked, significantly influences the system's performance. Its diameter, material, and length affect the pressure drop within the loop, impacting the pump's ability to circulate the coolant efficiently. Kinks and bends in the tubing increase the resistance, potentially reducing the cooling effectiveness. Material selection is crucial to avoid corrosion and compatibility issues with the coolant. Flexible tubing offers ease of installation, while rigid tubing can be better suited for high-pressure applications.

Radiators are the heat exchangers in liquid cooling systems, dissipating heat from the coolant to the ambient environment. Their surface area, fin density, and material significantly impact their performance. Larger surface areas generally lead to better heat dissipation but come with increased size and weight. Fin density is crucial; high density enhances heat transfer but also increases air resistance. Material selection similarly impacts heat transfer characteristics; aluminum and copper are common choices. Radiator size must be tailored to the heat load to ensure adequate cooling.

Beyond these core components, various other elements influence system performance. These include thermal interface materials (TIMs) that fill the gaps between the heat source and the heatsink, ensuring efficient heat transfer; sensors and controllers that regulate the system's operation and maintain optimal temperatures; and enclosures, which impact airflow and heat dissipation. Carefully considering these less prominent components is necessary for a well-functioning cooling system. Appropriate selection across all these components guarantees optimal performance.

3.2 Thermal Modeling

Thermal modeling is crucial for designing effective cooling systems. It allows engineers to predict the temperature distribution within a system under various operating conditions, before a physical prototype is built, saving time and resources. This predictive capability is paramount in ensuring the device operates within safe temperature limits and achieves its desired performance. Ignoring thermal modeling can lead to overheating, system failure, and even safety hazards.

Accurate thermal modeling relies on a thorough understanding of the heat transfer mechanisms at play. Conduction, convection, and radiation all contribute to the overall heat flow within the system. The model must accurately represent these processes, considering factors such as material properties, geometry, and boundary conditions. Software packages employing techniques like finite element analysis (FEA) and computational fluid dynamics (CFD) are commonly used for detailed simulations. These sophisticated tools enable the accurate prediction of temperature profiles, heat fluxes, and thermal stresses.

Choosing the right software depends on the complexity of the system. Simpler systems might be adequately modeled with readily available software, offering a user-friendly interface and quick results. More complex designs, however, might demand dedicated FEA or CFD packages capable of handling intricate geometries and boundary conditions. Regardless of the software, inputting the correct physical properties of materials, such as thermal conductivity and specific heat capacity, is crucial for realistic results. Using inaccurate values, even slight variations, will compromise the accuracy of the model.

Data validation is critical for effective thermal modeling. The model's predictions should be compared to experimental data or results from simpler models to ensure accuracy. Discrepancies between the model and the real-world performance indicate the need for adjustments to the model's parameters or the modeling methodology itself. Iterative refinement is key to improving the model's fidelity. This process ensures confidence in the model's predictions and allows for informed design decisions based on reliable data. Furthermore, validation can reveal unforeseen complexities or potential problems in the design that could not have been otherwise detected.

Beyond simply predicting temperatures, thermal modeling assists in optimizing the cooling system's design. By systematically altering design parameters within the model, engineers can evaluate the impact on overall thermal performance. For instance, changes to heat sink size, fan speed, or coolant flow rate can be easily simulated and their consequences analyzed. This iterative optimization process leads to a more efficient and cost-effective cooling solution, minimizing the need for extensive and costly prototyping. In short, simulation streamlines development by offering a virtual testing ground.

Effective thermal modeling also helps in identifying potential hotspots within the system. Hotspots are areas where the temperature exceeds safe operating limits, potentially leading to component damage or failure. The model highlights these areas, allowing engineers to take targeted corrective actions, such as adding extra cooling to those specific regions. Identifying and mitigating these hotspots is essential for ensuring the reliable long-term operation of the device. Without this crucial step, undetected hotspots could lead to catastrophic system failures.

The complexity of thermal modeling scales with the system's intricacy. A simple electronic component might only require a basic model accounting for heat conduction and convection. A high-power processor, however, necessitates a far more sophisticated model incorporating 3D heat transfer, fluid flow, and potentially radiation effects. The appropriate level of detail is crucial; oversimplification can result in inaccurate predictions, while excessive complexity can be computationally expensive and time-consuming. Finding the right balance between accuracy and computational efficiency is a key challenge.

Finally, remember that thermal modeling is an iterative process. The first model is rarely perfect. Expect to refine your approach based on experimental data and the insights gleaned from simulation results. This continuous improvement cycle is essential to developing a robust and reliable cooling system. Continual assessment and refinement ensure accuracy and optimal performance, enhancing the overall efficiency and lifespan of the cooled device. The use of thermal modeling is not a one-time event, but an integral part of a design methodology that ensures success.

3.3 Design Optimization

Effective cooling system design hinges on optimizing several interconnected factors. Prioritizing these elements ensures not only the efficient removal of heat but also the longevity and optimal performance of the cooled device. Careful consideration of each aspect, from component selection to the final testing phase, will yield a robust and reliable system. Neglecting any of these key areas can lead to suboptimal performance or even system failure.

Thermal modeling is your indispensable ally in this process. Before committing to any design, simulating heat transfer is crucial. Software packages, capable of performing sophisticated Finite Element Analysis (FEA), are available for accurately predicting temperature distributions within the system. By utilizing these tools, you can identify potential hotspots and design flaws early on, allowing for iterative refinements before expensive prototype fabrication. Don't underestimate the power of predictive modeling; it's a cost-effective means to avert costly mistakes down the line. Employing it wisely is paramount for a successful outcome. Moreover, understanding the limitations of these models and knowing when to validate the simulations with physical experiments is equally important.

Component selection is another critical step. The choice of heat sink material, for example, directly impacts the system's overall efficiency. Materials with high thermal conductivity, like copper or aluminum, are preferred for maximizing heat transfer. However, other properties such as cost, weight, and corrosion resistance must be considered. Similarly, fans must be carefully selected based on their airflow rate, noise level, and power consumption. A balance needs to be struck between performance and energy efficiency, and factors like system constraints and the operating environment must dictate the ultimate selections. The interaction between various components must be thoroughly considered; a well-designed heat sink is pointless if the airflow is inadequate.

Optimization often involves iterative refinement. After initial design and modeling, you may find areas that require improvement. Perhaps the heat sink isn't adequately transferring heat, or the airflow is poorly distributed. Modifying design parameters and repeating the modeling process allows for continuous improvement. This cyclical process, of analysis, modification and re-analysis, will yield optimal performance. This iterative approach underscores the need for flexibility and an understanding of how different design choices influence the outcome.

Testing and validation are essential final steps to confirm design effectiveness. Creating a prototype and subjecting it to controlled testing conditions, under various loads, is vital. This should include measuring temperatures at key locations within the system and comparing them against the predictions from the thermal models. Any discrepancies should trigger a review of the design process, highlighting aspects needing further

refinement. The testing and validation phase often reveal hidden issues not detectable through simulation alone. Thorough testing is the only way to ensure your system meets specifications in real-world conditions. Cost analysis is often overlooked, but its significance shouldn't be underestimated. While designing a highly efficient cooling system is important, it must be economically viable. Comparing the cost of different components, manufacturing processes, and potential energy savings throughout the system's lifetime is critical for any project. It is crucial to carefully evaluate the total cost of ownership (TCO), taking into account operational expenses over time. Finding the optimal balance between performance, efficiency and cost requires a nuanced approach and strategic planning. The most technically brilliant design is useless if it's financially unsustainable. Therefore, meticulous cost analysis is a critical component of the design optimization process.
Remember that design optimization is an iterative process. It's not a one-time event; instead, it requires a cyclical approach incorporating design, modeling, testing, and refinement. Throughout this continuous improvement process, your ultimate goal is to create a cooling system that efficiently dissipates heat, ensures reliable device operation, and optimizes performance while adhering to budget constraints. A well-designed system will provide the optimal balance between efficiency and cost-effectiveness.

3.4 Testing and Validation

Testing and validation are crucial; they're not mere afterthoughts, but integral parts of the design process. A well-designed cooling system, on paper, might fail spectacularly in the real world without rigorous testing. This isn't just about proving your design works; it's about identifying weaknesses, refining performance, and ensuring the longevity of your system. Ignoring this stage invites costly failures and compromises the reliability of your entire project.

Consider your specific device and its operating environment. What are the anticipated thermal loads? Will the device operate continuously or intermittently? These factors dictate the necessary testing parameters and conditions. Don't underestimate the importance of environmental factors—ambient temperature fluctuations, airflow restrictions, and even dust accumulation can significantly affect performance. Think about edge cases; simulating extreme conditions helps ensure your cooling system can handle unforeseen challenges.

We need to establish clear metrics for success. What constitutes acceptable operating temperatures? What level of heat dissipation is required? Setting these targets beforehand provides benchmarks for evaluating performance during testing. These metrics should align with the device's specifications and the overall system requirements. Using standardized testing procedures ensures consistency and reproducibility, enabling comparisons across different designs and iterations.

Your testing methodology must be comprehensive. Initial bench-top testing offers a controlled environment for evaluating individual components and subsystems. This allows you to isolate potential issues and address them before integrating everything into the final system. Thermocouples are your friends here, precisely measuring temperatures at various points to pinpoint thermal bottlenecks. Next, integrate testing simulates the real-world operational environment, subjecting the entire cooling system to realistic conditions. Monitoring key parameters like temperature, airflow, and pressure will reveal areas for optimization.

After rigorous testing comes validation, comparing results to your initial design specifications and the set benchmarks. Does your cooling system meet the established performance targets? If not, why not? Identifying discrepancies requires careful examination, possibly necessitating design revisions or component replacements. Data logging, meticulous record-keeping, and thorough analysis are crucial here, forming the basis for informed decision-making.

Remember, validation is an iterative process. Testing may reveal unexpected behaviors requiring further refinement. You may need to iterate the design, retest, and revalidate until the system consistently performs to the set specifications. This iterative loop is central to the development of a truly robust and reliable cooling solution. This is not a linear process. Expect to go back and forth, refining the system until it meets all requirements and passes your tests with flying colors.

Let's talk specifics. For instance, when testing a liquid cooling system, you might focus on the pump's efficiency, pressure drop across the system, and the coolant's temperature stability. The data collected is critical, but you also must pay attention to potential issues like leaks or corrosion. Careful examination of these issues could save you immense time and prevent costly repairs later. For air-cooled systems, air flow patterns and the effectiveness of the heat sink are crucial areas of focus. Detailed measurements of airflow and temperature gradients around the heat sink are necessary to optimize its design and placement.

Moreover, consider long-term testing. Continuous operation over extended periods reveals issues that might not surface in shorter tests. This is where you might observe issues like thermal fatigue in components or gradual degradation in performance. Long-term testing often highlights subtle issues. These subtle issues can become significant problems over time, possibly leading to premature component failure. Such testing adds another layer of assurance, providing a strong basis for confident deployment.

Finally, documentation is crucial. Comprehensive reports, including detailed test procedures, raw data, and analysis of results, provide a valuable record of your validation process. These documents are not only essential for internal reviews but also provide a robust foundation for future improvements. Furthermore, your records can support troubleshooting efforts, which are almost always needed. Well-maintained documentation is a crucial investment in the long-term success of your project. It also supports any future endeavors, offering invaluable knowledge for subsequent designs.

3.5 Cost Analysis

Cost-effective cooling solutions are paramount for ensuring optimal device functionality and longevity. Ignoring cost analysis during the design phase can lead to unforeseen budget overruns, delaying project completion, and ultimately compromising the system's performance. A thorough cost analysis should be integrated into every stage of the design process, from initial concept to final implementation.

Understanding the total cost of ownership (TCO) is vital. This goes beyond the initial capital expenditure (CAPEX) on components like pumps, heat sinks, and refrigerants. It includes operational expenses (OPEX) such as energy consumption, maintenance, and potential replacements over the system's lifespan. A detailed breakdown of these costs allows for informed decision-making and the selection of the most economically viable cooling strategy.

Component selection significantly impacts the overall cost. While high-performance components might offer superior cooling capabilities, they often come with a higher price tag. It's essential to find the optimal balance between performance and cost by comparing various options and evaluating their price-to-performance ratio. For instance, a larger, more powerful fan might provide better airflow but could increase energy consumption and initial cost, ultimately negating potential savings. Careful consideration is needed.

Thermal modeling plays a crucial role in cost optimization. Accurate simulations allow for the prediction of heat transfer rates and temperature distributions, enabling designers to identify potential hotspots and optimize component placement for maximum efficiency. This allows for the use of more cost-effective components without sacrificing performance, reducing the overall cost while ensuring optimal device operation. Over-engineering a system with excessive cooling capacity only drives up costs unnecessarily.

The cost analysis also encompasses indirect expenses. These include labor costs for installation, maintenance, and potential repairs. Factors such as downtime, productivity loss, and the cost of specialized tools or expertise should be included in the overall cost assessment. A poorly designed system, despite a low initial cost, might lead to expensive repairs and prolonged downtime, making it economically unfavorable in the long term.

Furthermore, the environmental impact of the cooling system should be considered, impacting the long-term cost. Energy-efficient cooling solutions reduce operational costs and minimize the environmental footprint, making them more sustainable and ultimately cost-effective in the long run. Using eco-friendly refrigerants or employing passive cooling techniques whenever feasible can reduce both direct and indirect costs associated with energy use and waste disposal.

Finally, the cost analysis should not be a standalone exercise. It should be an iterative process integrated into the entire design process. By continuously evaluating the cost implications of design choices, engineers can refine their strategies to optimize both performance and affordability. This approach ensures the system achieves its intended functionality while remaining within budget constraints, making it a sustainable and economically sound solution. This proactive approach prevents costly modifications or rework later in the process.

4 Air Cooling Techniques

Air cooling, a cornerstone of thermal management, relies on the convective transfer of heat from a device to the surrounding air. Understanding this process is paramount; it involves the movement of air, facilitated by fans, across heat-generating components, drawing away heat and maintaining operational temperatures within acceptable limits. Effective air cooling hinges on optimizing this heat transfer, minimizing thermal resistance, and ensuring efficient airflow. The choice of fan and heat sink is critical and needs careful consideration.

Heat sinks, passive components, augment the cooling process by providing an increased surface area for heat dissipation. The design of a heat sink directly impacts its effectiveness. Factors such as fin geometry, material, and surface treatment significantly influence the heat transfer rate. A larger surface area generally translates to better cooling performance, but also increases weight and cost. Careful consideration of these tradeoffs is essential in selecting the appropriate heat sink. Material selection, whether aluminum or copper, affects thermal conductivity and thus cooling efficiency.

Fans, the active component in air cooling, create airflow, driving the convective process. Fan characteristics, such as airflow rate (CFM), static pressure, and noise level, are critical considerations in system design. High CFM fans move larger volumes of air, effectively cooling hotter components. However, a higher CFM rating doesn't automatically equate to better cooling. Static pressure, the ability to overcome airflow resistance, is crucial when cooling densely packed components within enclosed systems. A balance must be struck between the desired cooling capacity and the acceptable noise level. Choosing the right fan is a vital step.

Effective airflow management is just as crucial as choosing the right fans and heat sinks. Improper airflow patterns can drastically reduce the efficiency of a cooling system. Ducting, strategic placement of components, and consideration of internal air channels can significantly enhance cooling. A well-designed system guides the airflow directly onto heat-generating components, maximizing heat removal. Obstacles to airflow, such as obstructions or poorly placed components, should be minimized. Computational fluid dynamics (CFD) simulations can be invaluable tools in optimizing airflow patterns. Using these tools ensures optimal cooling before the actual prototype creation.

The interplay between the fan and heat sink must be well-defined. A high-performance fan may be ineffective if paired with a poorly designed heat sink unable to dissipate the heat rapidly. Conversely, an excellent heat sink will be underutilized without sufficient airflow. Therefore, selecting these components requires a careful understanding of their individual characteristics and how they interact within the entire cooling system. Matching the fan's airflow and the heat sink's capacity is crucial for effective cooling.

Beyond component selection, proper system assembly is critical for optimal performance. Ensure components are properly secured and mounted to maximize thermal contact. The use of thermal interface materials, such as thermal paste or pads, is essential to minimize thermal resistance between the heat-generating component and the heat sink. Improper application of these materials can severely impede heat transfer. This step is essential to achieve the best possible thermal transfer.

Finally, monitoring and maintenance are key aspects of prolonging the lifespan and efficiency of air cooling systems. Regular cleaning of components, such as fans and heat sinks, is necessary to remove dust buildup. Dust accumulation significantly reduces airflow and thermal transfer effectiveness. Monitoring operating temperatures through sensors is crucial to detect potential problems early on, preventing overheating and system failure. By understanding and incorporating these techniques, engineers and designers can build reliable, efficient, and cost-effective air cooling systems.

4.1 Fans and Heat Sinks

Effective heat dissipation is crucial for maintaining optimal device functionality and longevity. Fans and heat sinks, often working in tandem, are fundamental components in many air cooling systems. Understanding their individual roles and how they interact is key to designing efficient cooling solutions.

The primary function of a fan is to create airflow. This airflow, acting as a convection medium, carries away heat generated by the device. Fan selection involves careful consideration of several key parameters. Airflow rate, measured in cubic feet per minute (CFM), is a critical specification, directly impacting the amount of heat that can be removed. Static pressure, the ability of the fan to overcome resistance within the cooling system, is equally important. Higher static pressure is necessary when dealing with restricted airflow pathways, such as densely packed components within a chassis. Choosing a fan with inadequate static pressure might result in insufficient cooling, leading to overheating. Noise levels are another crucial factor; a balance must be struck between cooling performance and acceptable noise emission. Finally, power consumption should be carefully considered, especially for portable devices or systems with limited power budgets. A highly efficient fan minimizes energy waste, contributing to overall system efficiency.

Heat sinks, passive components, augment the cooling process by increasing the surface area available for heat transfer. The larger surface area allows for greater convective heat exchange with the surrounding air. Heat sinks are typically constructed from materials with high thermal conductivity, such as aluminum or copper, to efficiently transfer heat from the device to their fins. The design of a heat sink's fin structure is critical to its effectiveness. Fin density, height, and spacing all influence its thermal performance. Denser fins increase surface area but can also impede airflow, highlighting the importance of a balanced design. Heatsink selection must carefully consider the heat load generated by the device and the ambient temperature conditions. An inadequately sized heat sink will fail to dissipate the generated heat, while an oversized one might be unnecessarily expensive and bulky.

The synergistic relationship between fans and heat sinks is crucial for effective cooling. A well-designed system carefully integrates these two components. Fans provide the airflow, while heat sinks maximize heat transfer to that airflow. The spacing and arrangement of heat sinks relative to the fan are significant factors. Proper airflow management ensures that the air efficiently sweeps across the heat sinks, maximizing heat removal. Poorly designed airflow pathways, such as obstructed paths or uneven air distribution, can drastically reduce overall system performance. The optimization process often involves iterative design, simulation, and testing. Thermal modeling software can assist in predicting the performance of various configurations.

The selection of fans and heat sinks must always consider the specific application and environmental conditions. High-performance computing systems, for instance, demand more powerful fans and larger heat sinks than embedded systems with lower heat dissipation requirements. In addition, the ambient temperature plays a significant role. Higher ambient temperatures reduce the effectiveness of air cooling, necessitating more robust components.

Ultimately, choosing appropriate fans and heat sinks involves a balance between performance, cost, and size constraints. Carefully considering the parameters described above is essential for creating an effective cooling solution. Overlooking these aspects can lead to system instability, premature component failure, and ultimately, system failure. Thorough analysis and design optimization, including thermal modeling and validation, are vital to ensure that the chosen cooling system meets the performance requirements while minimizing cost and physical footprint. The interaction between the fan, producing airflow, and the heat sink, maximizing surface area for heat transfer, underscores the need for a holistic approach to cooling system design. Effective cooling is not just about individual component performance, but the synergistic interplay between them.

4.2 Airflow Management

Effective airflow management is paramount in optimizing the performance and longevity of any air-cooled system. Poor airflow can lead to overheating, component failure, and reduced lifespan, negating the benefits of even the most advanced cooling components. Therefore, understanding and implementing proper airflow techniques is crucial.

Strategic placement of components within the system is the first step towards effective airflow management. Components that generate the most heat should be positioned to maximize exposure to the cooling air stream. This often involves careful consideration of the overall system layout and the direction of the airflow, ensuring that hot air is efficiently carried away from heat-generating elements. Avoid obstructing airflow paths with unnecessary components or cabling. Efficient spacing between components helps to reduce thermal hotspots and promote better overall heat dissipation.

The design of air ducts and channels significantly impacts airflow efficiency. The shape, size, and arrangement of these channels influence air pressure, velocity, and turbulence. Designing them to minimize resistance and maintain consistent airflow throughout the system is critical. This will allow for consistent cooling throughout the device. Sharp bends and obstructions should be avoided as much as possible because they create pressure drops and turbulence which reduces the effectiveness of air cooling. Smooth transitions and strategically placed baffles can help in minimizing these negative effects and maintaining a controlled flow.

Fan selection plays a crucial role in airflow management, particularly the fan's static pressure. Higher static pressure fans are often preferred for systems with high airflow resistance, such as those with densely packed components or intricate ductwork. They provide the necessary pressure to overcome such resistance and ensure adequate airflow. However, for systems with less restrictive airflow paths, lower static pressure fans can be more energy-efficient without sacrificing cooling effectiveness. Choosing the right fan for the application is essential for maximizing efficiency and minimizing noise. In some cases, multiple fans might be necessary for cooling a larger device, thereby enhancing performance and device lifespan.

Heat sinks are vital components in any air cooling system and their design directly influences heat dissipation efficiency. Properly designed heat sinks increase the surface area available for heat transfer, thereby aiding convection. The fin geometry, material, and contact area with the heat source are important parameters affecting thermal performance. The choice of the material used to construct the heat sinks also impacts efficiency. Consider materials with high thermal conductivity, such as copper or aluminum, to ensure effective heat transfer. These metals can improve heat dissipation and reduce overall system temperatures.

The efficiency of an air cooling system can be improved through the proper usage of air filters. Filters prevent the accumulation of dust and debris on heat sinks and other components, ensuring that the airflow is not significantly reduced over time. Regular cleaning or replacement of air filters is a simple and inexpensive measure to ensure that the airflow is as efficient as possible.

Beyond hardware, software tools can significantly enhance airflow management. Computational fluid dynamics (CFD) simulations provide a powerful way to model and visualize airflow patterns within a system.

CFD allows for optimizing ductwork design, fan placement, and component arrangement. This analysis of the airflow helps in predicting temperature distributions and identifying potential hot spots. This predictive capability enables preemptive adjustments and mitigates potential failures. This detailed analysis allows engineers to make informed decisions that lead to optimal cooling performance.

Finally, thorough testing and validation are essential for verifying the effectiveness of the implemented airflow management techniques. This involves measuring temperatures at various points within the system under different operating conditions. The collected data should be analyzed to identify any areas that require improvement. This process often involves iterative refinement of the design to achieve the desired cooling performance. Careful monitoring is critical for ensuring that the system continues to operate efficiently over time. Regular thermal monitoring and occasional system cleaning are simple but very effective steps for maintaining optimal cooling and preventing device degradation. By paying attention to airflow, system life can be extended and improved.

5 Liquid Cooling Methods

Water cooling, a cornerstone of advanced thermal management, offers significantly improved heat dissipation compared to air cooling. Its effectiveness stems from the high specific heat capacity of water, allowing it to absorb and transport substantial amounts of heat energy. Understanding the fundamental principles–pump selection, proper tubing configuration, and radiator design–is paramount for building a robust and efficient system. The choice of pump, whether a centrifugal, peristaltic, or other type, greatly impacts flow rate and pressure, both crucial for optimal cooling. Tubing material, internal diameter, and length all influence flow resistance and, consequently, the system's overall efficiency. Radiators, acting as heat exchangers, must possess sufficient surface area to effectively transfer heat from the liquid coolant to the surrounding environment. Careful consideration of these interconnected components is essential for achieving the desired cooling performance.

Beyond the basics, advanced liquid cooling techniques push the boundaries of thermal management. Microchannel cooling, for example, employs intricate networks of miniature channels etched into a substrate. This dramatically increases the surface area available for heat transfer, resulting in significantly higher cooling capacities. Microfluidic systems leverage precise control over fluid flow at the microscopic level, enabling even more efficient heat removal. These advanced techniques are particularly crucial in high-power density applications, such as high-performance computing and electronics, where conventional methods might prove insufficient. The complexity of designing and manufacturing these systems demands specialized expertise and precision engineering. However, the performance gains often justify the added investment.

Refrigerant cooling represents a distinct approach to liquid cooling, utilizing fluids with superior thermodynamic properties for heat transfer. These refrigerants, often characterized by low boiling points and high latent heats of vaporization, are particularly well-suited for applications requiring extremely efficient cooling. Refrigerant systems typically incorporate a compressor, condenser, expansion valve, and evaporator, creating a closed-loop cycle that continuously circulates the refrigerant. Selecting the appropriate refrigerant requires careful consideration of its thermodynamic properties, environmental impact, and safety considerations. The design of the system's components–the compressor's efficiency, condenser size, and evaporator design–all significantly influence overall cooling effectiveness. While providing superior cooling performance, refrigerant systems often necessitate more intricate designs and specialized components, increasing their complexity and cost compared to simple water cooling. The system's efficiency is determined by the interplay of several factors, including the refrigerant's properties, component design, and operating conditions. Careful design and optimization are crucial for maximizing its efficiency and minimizing environmental impact. The selection of components necessitates a careful balance between performance and cost. Specialized expertise is often required for efficient design, implementation, and maintenance of refrigerant cooling systems, especially in demanding applications. The careful selection and precise control of refrigerant flow rates and pressures are critical for maintaining optimal performance and preventing any equipment malfunctions. Regular system maintenance is crucial to ensure long-term reliability and efficiency. Advancements in refrigerant technology continue to improve the efficiency and environmental friendliness of these systems. Further research focuses on developing alternative refrigerants with reduced global warming potentials, striving for enhanced environmental sustainability.

Proper system design requires a multifaceted approach, balancing component selection with the overall system's requirements. The effectiveness of liquid cooling hinges on careful consideration of multiple interacting factors; a poorly designed system can lead to reduced efficiency or even equipment failure. The principles of fluid mechanics, heat transfer, and thermodynamics are fundamental to achieving optimal performance. A thorough understanding of pump characteristics, pressure drops within the system, and heat transfer coefficients is essential. Moreover, appropriate considerations for material compatibility, corrosion resistance, and system safety are vital to long-term reliability and operational longevity. The choice of materials, such as those used in tubing and components that come into direct contact with the cooling liquid, is crucial to ensuring both functionality and long-term durability. Corrosion and biofouling can significantly reduce the efficiency and lifespan of a liquid cooling system. Therefore, selecting appropriate materials and implementing proper maintenance strategies are crucial for reliable operation. Ultimately, effective liquid cooling involves integrating multiple engineering disciplines to achieve a harmonized and reliable system. Careful planning and selection of components and materials are vital for achieving the desired cooling performance and ensuring the longevity of the system.

5.1 Water Cooling Basics

Water cooling offers a significant advantage over air cooling for high-heat-generating components, providing superior heat dissipation capabilities. The fundamental principle involves circulating a liquid coolant, typically water, through a system to absorb heat from the heat source, then dissipating this heat via a radiator. This continuous circulation maintains a lower operating temperature, preventing thermal throttling and extending component lifespan. Understanding the components and their interplay is crucial for building an effective water cooling loop.

A typical water cooling system includes several key elements: a water block, which directly contacts the heat-generating component; a pump, responsible for circulating the coolant; tubing, which connects the various parts; a radiator, where heat is transferred to the surrounding air; and a reservoir, often integrated into the pump or a separate unit, to hold the coolant. Proper selection of these components based on the heat load and desired performance is critical for optimal operation. For example, a high-performance CPU might necessitate a larger radiator and a more powerful pump compared to a system cooling a less demanding component.

The choice of coolant is also a significant factor. While distilled water is commonly used, adding a coolant additive is strongly recommended. These additives prevent corrosion, algae growth, and scale formation within the loop, ensuring long-term system stability and performance.

Choosing a high-quality additive that's compatible with the system materials is vital for the longevity of the entire setup. Regular maintenance, including fluid changes and cleaning, is crucial to prevent the accumulation of contaminants and maintain optimal heat transfer.

Pump selection is another critical aspect of water cooling design. Choosing an appropriate pump ensures sufficient coolant flow to effectively remove heat from the component. Factors to consider include flow rate, pressure head, and noise level. High-flow pumps are ideal for larger systems or those with high heat loads, while quieter pumps might be preferred for systems where noise is a concern. The pump's placement within the loop also influences performance. Optimal pump placement minimizes pressure drop and ensures even coolant distribution.

Tubing material is important, too. While several materials exist, each has its own advantages and disadvantages. Some common materials are rubber or PVC, which are relatively inexpensive, and harder tubing made of acrylic or PETG. It is essential to select tubing that is both durable and compatible with the coolant and additives used. Proper bending techniques are necessary to prevent kinks or restrictions that could impede flow and compromise cooling performance. Secure connections are absolutely critical to prevent leaks.

Radiator selection is crucial for efficient heat dissipation. Larger radiators, with more surface area, provide improved cooling capacity. The radiator's fin density, material and design significantly influence its effectiveness. Radiator placement should ensure adequate airflow, maximizing convective heat transfer to the ambient environment. Proper mounting and airflow management are crucial for achieving optimal heat dissipation from the radiator, which is essential for keeping the entire system cool. Incorrect installation could lead to significant performance penalties.

Finally, ensuring proper system assembly and leak testing is critical before operation. A poorly assembled system can lead to leaks, which can damage components and create safety hazards. Thorough testing and pressure testing is essential to identify and resolve any leaks before starting the loop. Regular monitoring of the system's performance, including coolant temperature and flow rate, helps in identifying potential issues early on, and should be considered standard practice. Maintenance, such as regular fluid changes and cleaning, is crucial for ensuring that the cooling system continues to operate at peak efficiency over its lifespan.

5.2 Advanced Liquid Cooling

Advanced liquid cooling surpasses basic water cooling by employing sophisticated techniques and components to achieve superior heat dissipation. Microchannel cooling, for instance, utilizes intricate networks of tiny channels etched into a solid substrate, maximizing the surface area available for heat transfer. This significantly enhances the coolant's ability to absorb heat from the heat source, leading to lower operating temperatures and improved performance. The design intricacies involve careful consideration of channel dimensions, flow rates, and fluid properties to optimize heat transfer coefficients. Precise manufacturing techniques are crucial for creating these microchannels, demanding high precision and sophisticated fabrication processes. Materials selection also plays a vital role; materials with high thermal conductivity, corrosion resistance, and biocompatibility are essential for long-term reliability and optimal performance.

Unlike basic water cooling loops, advanced systems frequently incorporate sophisticated pump technologies to maintain consistent and high-velocity coolant flow through these intricate microchannels. High-performance pumps are necessary to overcome the increased frictional losses associated with the smaller channel dimensions. These pumps often employ advanced designs, such as centrifugal or peristaltic pumps, that offer superior efficiency and pressure capabilities, ensuring uniform cooling across the entire heat-generating surface. Furthermore, the incorporation of variable speed control allows for dynamic adjustment of the flow rate according to the heat load, optimizing energy consumption and maintaining optimal operating temperature.

Choosing the correct coolant is paramount. While water is often used in basic systems, advanced applications may require specialized coolants with enhanced thermal properties. Fluids such as engineered coolants with additives that improve heat transfer characteristics, corrosion inhibition, and anti-fouling properties are frequently used. These specialized fluids offer significant advantages over water, especially in high-performance applications where elevated temperatures or demanding operating conditions might degrade the performance of ordinary water. Careful consideration of coolant compatibility with the system materials is absolutely crucial to prevent corrosion and maintain system integrity. Regular fluid analysis and replacement are often necessary to maintain the coolant's efficiency and prevent fouling or degradation.

System monitoring and control are also significantly more advanced in these sophisticated setups. Sensors strategically placed throughout the cooling loop provide real-time data on temperature, flow rate, and pressure. This information feeds into a control system that dynamically adjusts pump speed, coolant flow distribution, and other parameters to maintain optimal operating conditions. This closed-loop control provides improved stability and responsiveness compared to simpler systems, enhancing the overall reliability and performance of the system. Sophisticated control algorithms can predict potential issues and proactively adjust parameters to prevent overheating or system failures, adding a new layer of safety.

Beyond microchannels, immersion cooling represents another advanced liquid cooling technique. Here, the heat-generating component is completely submerged in a dielectric fluid, often a specialized coolant with high thermal conductivity and dielectric strength. This method offers several advantages, including direct contact cooling, eliminating the thermal resistance associated with interfaces between the heat source and the coolant. The effectiveness hinges on the fluid's ability to rapidly absorb heat and its ability to remain stable under high temperatures and pressures. Proper fluid selection based on the specific application (e.g., considering flammability, toxicity, and environmental impact) is vital for ensuring both safety and efficacy. Furthermore, selecting the appropriate container material, considering its thermal conductivity and compatibility with the coolant, directly influences the overall efficiency of the immersion cooling system.

Finally, the integration of advanced liquid cooling systems often necessitates a meticulous approach to thermal management. This includes careful consideration of heat dissipation pathways, minimizing thermal resistances, and ensuring efficient heat removal from the coolant to the ambient environment. Sophisticated techniques such as optimizing component placement, utilizing high-performance heat exchangers, and employing advanced thermal modeling are employed. Thermal simulation using sophisticated software plays a crucial role in optimizing the design and predicting the performance of the system. Rigorous testing and validation are also essential to ensure the system meets performance requirements and operates reliably under real-world conditions. The integration of such advanced cooling solutions often demands a multifaceted, interdisciplinary approach, involving engineers from diverse backgrounds and expertise.

5.3 Refrigerant Cooling

Refrigerant cooling utilizes the thermodynamic properties of refrigerants to achieve efficient heat transfer. Unlike air or water cooling, which rely on sensible heat transfer, refrigerant systems leverage the latent heat associated with phase changes–evaporation and condensation–resulting in significantly higher cooling capacity per unit volume. This makes refrigerant cooling particularly valuable for applications requiring high heat dissipation, such as in data centers, industrial chillers, and high-performance computing systems. The selection of the appropriate refrigerant is critical, influenced by factors such as environmental impact (ozone depletion potential and global warming potential), thermodynamic properties (boiling point, critical temperature, and latent heat of vaporization), and safety considerations (flammability, toxicity).

The basic principle involves a closed-loop system where the refrigerant circulates continuously. A compressor raises the refrigerant's pressure and temperature, forcing it into a condenser where it releases heat to the surroundings, typically through air or water cooling, and condenses back into a liquid. This liquid refrigerant then flows through an expansion valve, causing a significant pressure drop and subsequent temperature reduction. This low-temperature, low-pressure refrigerant then absorbs heat from the device being cooled in the evaporator, evaporating and transitioning back into a gaseous state. This cycle repeats, providing continuous cooling. Understanding the pressure-temperature relationships of the chosen refrigerant is essential for optimizing system performance. Precise control of pressure and temperature is maintained through the use of thermostatic expansion valves or electronic expansion valves, ensuring efficient operation and preventing system inefficiencies.

Efficient operation necessitates careful consideration of component design and integration. The compressor, the heart of the system, requires selection based on cooling capacity, energy efficiency (measured in COP or coefficient of performance), and compatibility with the refrigerant. Condensers are typically designed to maximize heat transfer to the surroundings, often utilizing finned surfaces to enhance surface area and improve airflow or water flow. The evaporator, designed to maximize heat absorption from the cooled object, similarly requires attention to surface area and fluid flow characteristics. The expansion valve plays a crucial role in regulating refrigerant flow, controlling the pressure drop, and ultimately the temperature of the refrigerant entering the evaporator.

Proper design necessitates the use of appropriate materials resistant to corrosion and compatible with the refrigerant, considering factors such as pressure and temperature ranges. Furthermore, insulation plays a key role in minimizing heat leaks, thereby improving the system's overall efficiency and reducing energy consumption. The integration of the cooling system with the device or process it is designed to cool is critical to maximize effectiveness, requiring careful consideration of thermal contact resistance between the cooled object and the evaporator. A poorly designed interface can significantly diminish the system's overall performance. Proper system sizing based on the heat load of the device and ambient conditions is crucial to avoid oversizing or undersizing, both of which can negatively impact efficiency and cost-effectiveness.

Careful consideration of safety protocols is crucial throughout the design and operation phases. Refrigerants, depending on their chemical properties, may be flammable, toxic, or both. Leak detection systems and appropriate safety measures must be incorporated into the design to minimize risks. Regular maintenance, including checking for leaks and ensuring proper refrigerant levels, is critical for maintaining optimal performance and preventing potential hazards. The choice of refrigerant greatly influences safety considerations. Hydrofluoroolefins (HFOs) and hydrofluorocarbons (HFCs), while having lower ozone depletion potential than older refrigerants, still contribute to greenhouse gas emissions. The ongoing development of environmentally friendly refrigerants with low global warming potential is a significant area of ongoing research and development, constantly driving the evolution of refrigerant cooling systems. The overall performance of the system is best monitored and improved by regular system checks, and adjustments. The optimization process should always be a continuous effort in order to guarantee efficient and safe operation. Regular maintenance, including inspections for leaks, cleaning of components, and performance checks, ensures longevity and optimal efficiency, preventing premature failure and ensuring continuous cooling. This proactive approach not only improves the lifespan of the cooling system but also mitigates potential safety risks associated with refrigerant leaks.

6 Phase-Change Cooling

Phase-change cooling leverages the significant latent heat associated with phase transitions–boiling and condensation–to achieve highly efficient heat removal. Unlike sensible heat transfer, where temperature changes drive heat flow, phase change involves a substantial energy absorption or release without a corresponding temperature shift. This makes it exceptionally effective for managing high heat fluxes, common in modern electronics and other high-power applications. Consider, for instance, the boiling of water; a large amount of energy is absorbed during the transition from liquid to vapor, even though the temperature remains constant at 100°C (at standard atmospheric pressure). This principle underpins the effectiveness of phase-change cooling.

Immersion cooling, a prominent phase-change method, directly submerges the heat-generating component in a dielectric coolant. This ensures intimate contact between the heat source and the cooling fluid, minimizing thermal resistance and maximizing heat transfer. Common coolants include fluorocarbons, which possess excellent dielectric properties, preventing electrical shorts while efficiently absorbing heat. The boiling of the dielectric fluid around the component provides effective heat removal, while the vapor generated is managed by a condenser, typically located outside the immersion chamber, to maintain a stable cooling loop. Careful consideration of the coolant's properties, including its boiling point, thermal conductivity, and dielectric strength, is crucial for optimal system design and safety. The choice of coolant impacts the system's efficiency, operating temperature range, and overall cost.

Two-phase cooling, a more complex approach, utilizes both the liquid and vapor phases of a coolant simultaneously to enhance cooling capacity. In a typical setup, the coolant is circulated in a closed loop, where it absorbs heat in the evaporator section, transitioning to a two-phase mixture of liquid and vapor. This mixture then moves to a separator, which separates the vapor from the liquid. The vapor is subsequently condensed and returned to the liquid phase, completing the cycle. The advantage of two-phase cooling lies in its ability to handle significantly higher heat fluxes than single-phase systems, making it ideally suited for high-power density applications such as data centers and high-performance computing systems. However, the added complexity increases the system's cost and requires careful control of the two-phase flow to prevent issues such as flow instabilities or dry-out conditions.

Boiling, a crucial aspect of phase-change cooling, is a complex process influenced by factors like surface characteristics, coolant properties, and heat flux. Nucleate boiling, where bubbles form and depart from the heated surface, is highly desirable due to its enhanced heat transfer coefficient. However, as the heat flux increases, different boiling regimes can occur, including film boiling, where a vapor film insulates the surface from the liquid, leading to a decrease in heat transfer effectiveness. Understanding these boiling regimes is essential for designing efficient and reliable systems. Careful selection of the heating surface material and its surface treatment can significantly influence the boiling process, promoting nucleate boiling and delaying the onset of less desirable regimes. Condensation, the counterpart to boiling, is equally critical in phase-change cooling systems. It represents the transition of the coolant vapor back to the liquid phase, releasing the absorbed heat. The efficiency of the condensation process is influenced by factors such as surface area, coolant properties, and temperature difference between the vapor and the condensing surface. Enhanced surface designs, such as microchannel condensers, can significantly improve the condensation rate by increasing the surface area and promoting better contact between the vapor and the cooling surface. The proper design and implementation of the condensation process are crucial for maintaining the stability and efficiency of the entire cooling loop, ensuring consistent heat removal and preventing system overheating. Optimization techniques often involve using materials with high thermal conductivity and surface tension to enhance the condensation process and ensure optimal heat transfer. The condenser's design must be carefully chosen to match the operating conditions and avoid pressure drops that might negatively impact the overall system performance. Careful consideration of these aspects guarantees the effective and efficient operation of the phase-change cooling system.

6.1 Boiling and Condensation

Boiling and condensation represent a powerful duo in the realm of phase-change cooling. These processes harness the latent heat of vaporization and condensation, significantly boosting heat transfer compared to single-phase methods. Understanding the intricacies of these processes is crucial for effective system design. The large amount of energy absorbed during boiling and released during condensation provides a highly efficient mechanism for heat removal.

Effective boiling relies on several key factors. Surface properties, including roughness and wettability, significantly influence bubble nucleation and growth. A rough surface provides more nucleation sites, promoting more uniform boiling and preventing the formation of large vapor bubbles that can insulate the heating surface. Conversely, a smooth, non-wetting surface can lead to film boiling, a less efficient regime characterized by a vapor film separating the liquid from the surface. The choice of coolant also plays a pivotal role, as its thermophysical properties–like latent heat, thermal conductivity, and surface tension–directly affect boiling performance. Coolants with high latent heat and thermal conductivity are generally preferred, as they can absorb and transport more heat.

Condensation, the counterpart to boiling, is equally critical. It's the process where a vapor transforms into a liquid, releasing the latent heat of vaporization. This heat release needs to be efficiently managed to prevent overheating. Several factors influence condensation efficiency. Surface characteristics once again play a crucial role. A smooth, clean surface promotes filmwise condensation, where a continuous liquid film forms on the surface. While simple to understand, this can create a thermal resistance that limits heat transfer. Conversely, dropwise condensation, characterized by discrete droplets forming on the surface, generally offers superior heat transfer due to the increased surface area exposed for heat exchange. However, maintaining dropwise condensation often requires surface treatments or additives and can be challenging to achieve consistently. The selection of the condensing surface material also impacts the

efficiency of the process. Materials with high thermal conductivity will facilitate efficient heat transfer away from the condensation surface.
The design of a cooling system incorporating boiling and condensation often involves careful consideration of the heat transfer surface area. Increasing this area enhances heat transfer, but the design must balance efficiency with system size, cost, and other constraints. Many sophisticated methods exist, ranging from enhanced surfaces with micro\item or nano-structures to the use of porous materials or extended surfaces. The geometry of the system also plays a role; for instance, carefully designed channels can guide both the liquid and vapor phases, optimizing the contact area and overall performance.
Implementing boiling and condensation requires careful attention to several critical aspects to ensure stable and efficient operation. One key challenge is managing the flow patterns of the liquid and vapor phases, avoiding phenomena like dryout (where the heating surface becomes uncovered by liquid) which can lead to overheating and system failure. Appropriate flow control devices and design configurations need to be implemented to maintain adequate liquid supply to the boiling surface and efficient removal of vapor from the condensing surface. Additionally, pressure management is essential. The pressure within the system significantly affects boiling and condensation temperatures, and careful control is necessary to avoid unwanted effects.
Moreover, the selection of materials needs to consider their compatibility with the working fluids and the operating conditions. Corrosion and fouling can significantly impact the long-term performance of the system, and material choice is crucial to mitigate such problems. Finally, it's important to note the safety aspects of handling refrigerants, particularly

if dealing with high pressures or flammable coolants. Appropriate safety measures must be incorporated to protect personnel and equipment. A thorough understanding of the relevant regulations and safety protocols is crucial. Boiling and condensation offer a significant potential for efficient cooling, but practical implementation requires careful consideration of these critical design factors. The efficiency of these processes lies in optimizing surface properties, fluid choices, geometry, flow control, and pressure management. By carefully balancing these factors, engineers can effectively exploit the exceptional heat transfer capabilities of boiling and condensation, ultimately leading to superior device cooling solutions.

6.2 Immersion Cooling

Immersion cooling offers a compelling solution for high-power density electronics, surpassing traditional air and even some liquid cooling methods in its effectiveness. The fundamental principle involves submerging the heat-generating components–like CPUs, GPUs, or power electronics–directly into a dielectric fluid. This fluid, carefully selected for its thermal properties and electrical insulation capabilities, absorbs heat directly from the device's surface, drastically improving heat transfer efficiency. The elimination of air gaps and the high thermal conductivity of the dielectric fluid contribute significantly to enhanced cooling capabilities, enabling higher operational power and performance.

Unlike air cooling, which relies on convection currents to dissipate heat, immersion cooling ensures a consistent and direct contact between the heat source and the coolant. This direct contact minimizes thermal resistance, the primary impediment to efficient heat transfer. The large surface area of the submerged components, coupled with the fluid's high heat capacity, allows for rapid and uniform heat absorption, resulting in a significantly lower operating temperature compared to other cooling techniques. This lower operating temperature is crucial for preventing thermal throttling and extending the lifespan of the device. Imagine a scenario where thermal limits prevent your system from reaching its full potential; immersion cooling could be the solution to fully unlock that potential.

Choosing the right dielectric fluid is paramount. Factors such as its dielectric strength, thermal conductivity, viscosity, and boiling point all influence its suitability for a specific application. Fluids like perfluorocarbons (PFCs) or mineral oils are frequently utilized, each offering a different balance of properties. PFCs, for instance, boast exceptional dielectric strength and thermal conductivity, making them ideal for high-voltage applications and high-power density devices. However, their environmental impact and cost are significant factors to consider. Mineral oils present a more cost-effective alternative, but their dielectric strength might be lower, limiting their applicability in certain situations. A thorough analysis of these factors is essential to select the optimum fluid for a given system design.

The design of the immersion cooling system also plays a vital role in its effectiveness. Careful consideration must be given to factors such as tank size and shape, fluid circulation methods, and the placement of components within the tank to optimize heat transfer. For example, using strategically placed baffles or strategically shaping the tank can enhance fluid flow and promote uniform temperature distribution, preventing localized hot spots. Furthermore, the system should be designed to manage the expansion and contraction of the fluid as its temperature changes, preventing pressure build-up and potential leaks. Careful monitoring of the fluid's temperature and pressure is crucial to ensure safe and reliable operation of the entire system.

Beyond the choice of fluid and system design, the implementation of immersion cooling requires attention to safety precautions. The dielectric fluid, while non-conductive, should be handled with care, and proper safety measures should be implemented to prevent spills or exposure to personnel. Regular maintenance and inspection are crucial to identify potential issues and prevent unexpected malfunctions. This includes monitoring the fluid's cleanliness and regularly checking for leaks or damage to the system's components. Furthermore, proper grounding and isolation techniques are essential to ensure electrical safety, given the close proximity of the electronics to the liquid coolant.

In summary, immersion cooling presents a powerful and efficient method for managing heat in high-power density electronic devices. Its capacity to improve heat transfer significantly, leading to lower operating temperatures and extended lifespans, positions it as a technology worthy of careful consideration. However, successful implementation necessitates a meticulous approach to fluid selection, system design, and safety procedures, balancing cost, environmental impact and operational efficiency. Choosing the right dielectric fluid, designing for optimal fluid flow, and implementing stringent safety protocols are not merely suggestions, but rather essential steps toward the successful deployment and long-term reliability of an immersion cooling system. The complexities of immersion cooling demand a well-informed and proactive approach to avoid potential pitfalls and fully harness its considerable advantages. The potential benefits of prolonged device lifespan and increased operational performance often outweigh the initial investment and design complexities.

6.3 Two-Phase Cooling

Two-phase cooling leverages the substantial heat transfer that occurs during phase transitions–specifically, boiling and condensation. This process offers a significant advantage over single-phase methods because the latent heat of vaporization, the energy required to change a liquid into a gas, is considerably higher than the sensible heat, the energy needed to raise the temperature of a liquid. Harnessing this latent heat allows two-phase systems to absorb and remove far greater quantities of heat than equivalent single-phase systems, leading to dramatically improved cooling performance. Careful consideration of the operating parameters is vital for optimizing this efficiency.

The core principle behind two-phase cooling involves utilizing a refrigerant or coolant that undergoes a change in phase within the cooling system. For example, a liquid coolant absorbs heat from a device, evaporating into a vapor. This vapor, carrying away the absorbed heat, is then condensed back into a liquid, releasing the heat to the surroundings. This cyclical process continues, providing continuous cooling. The choice of refrigerant significantly impacts the system's performance and efficiency, with factors like boiling point, thermal conductivity, and environmental impact all playing crucial roles. Therefore, a thorough understanding of refrigerant properties is paramount for successful system design.

Designing a practical two-phase cooling system necessitates a keen understanding of thermodynamics and fluid mechanics. Factors such as pressure drop, flow rate, and heat transfer coefficients need meticulous evaluation and optimization. Accurate modeling of these processes, often employing computational fluid dynamics (CFD), is indispensable for predicting system behavior and identifying potential bottlenecks. For instance, correctly modeling the distribution of liquid and vapor phases within the system is critical for maximizing heat transfer efficiency. Ignoring these complexities can result in underperformance or even system failure.

Different configurations are employed in two-phase cooling systems, each with its own advantages and disadvantages. Immersion cooling, a commonly used technique, involves submerging the heat-generating device directly in a dielectric coolant. This method provides excellent heat transfer, especially for high-heat-flux applications. However, careful consideration must be given to the dielectric properties of the coolant to prevent electrical short circuits and ensuring compatibility with the device materials. The selection of an appropriate coolant, one with high thermal conductivity and sufficient dielectric strength, is thus a crucial design parameter. Another prevalent method is the use of two-phase loops, often employing a separate evaporator and condenser. The working fluid evaporates in the evaporator, absorbing heat from the heat source. The vapor is then transported to a condenser, where it releases heat and condenses back into a liquid. A pump circulates the liquid back to the evaporator, creating a closed loop. This approach provides greater flexibility in system design and allows for easier separation of the heat source and the heat rejection components. The design of the loop, including the choice of components like pumps, valves, and heat exchangers, is critical in achieving optimal efficiency and reliability.

The selection of appropriate materials is critical for both component longevity and overall system performance. Corrosion resistance is a particularly important consideration, especially in systems employing water or other aggressive coolants. The material choice should be compatible with the chosen coolant to prevent degradation and maintain system integrity over time. Furthermore, the thermal conductivity of materials utilized in the heat exchangers and other components directly influences the system's efficiency. Higher thermal conductivity materials enable more efficient heat transfer, leading to improved overall performance. This necessitates a careful trade-off between material properties, cost, and system requirements.

Maintaining optimal operating conditions is crucial for efficient and reliable two-phase cooling. This includes close monitoring of pressure, temperature, and flow rates. Sensors are often incorporated into the system to provide real-time feedback, which can be used to adjust operating parameters and prevent potential issues. Regular maintenance, such as cleaning and replacing components as needed, is also essential for long-term system performance and reliability. Ignoring these maintenance considerations can lead to decreased efficiency, increased risk of failure, and potential damage to the device being cooled.

In conclusion, two-phase cooling presents a powerful approach to managing heat in high-performance systems. However, the successful implementation requires a comprehensive understanding of thermodynamics, fluid dynamics, and materials science. Through meticulous design, material selection, and ongoing maintenance, engineers can leverage the unique advantages of two-phase cooling to achieve superior heat dissipation and ensure the continued performance and longevity of their cooled devices. Attention to detail at every stage is critical for maximizing efficiency, minimizing risks and ultimately ensuring long-term operational success.

7 Emerging Cooling Technologies

Thermoelectric cooling, a technology leveraging the Peltier effect, presents a compelling alternative to traditional methods. This effect, observed in junctions of dissimilar conductors, causes heat absorption at one junction and heat rejection at another when a direct current is applied. The temperature difference generated is directly proportional to the applied current, offering precise temperature control. Applications range from small-scale devices like CPU coolers to larger systems in specialized environments. However, the efficiency of thermoelectric coolers (TECs) is currently lower than that of vapor-compression systems, limiting widespread adoption. Ongoing research focuses on enhancing the performance of TECs through material science advancements, exploring novel materials with improved thermoelectric properties to increase efficiency and broaden applicability.

Advanced materials are revolutionizing the landscape of cooling system design. Nanofluids, suspensions of nanoparticles in base fluids, demonstrate significantly enhanced thermal conductivity compared to conventional coolants like water or ethylene glycol. This improvement stems from the high surface area of nanoparticles, which facilitates more efficient heat transfer. However, challenges remain in the stability and long-term performance of nanofluids. The potential for nanoparticle aggregation and clogging remains a key area of investigation, alongside concerns about the long-term environmental impact of the nanoparticles themselves. Carbon nanotubes, known for their exceptional thermal conductivity, are another promising material for heat sink applications. Their unique structure allows for efficient heat dissipation, making them ideal candidates for high-heat-flux devices. However, the cost of production and integration into existing systems remains a hurdle to widespread implementation. The development of effective, scalable manufacturing processes is critical for unlocking the full potential of carbon nanotubes in cooling applications.

The pursuit of improved heat transfer has led to innovative designs in heat sinks. Microchannel heat sinks, featuring intricate networks of tiny channels etched into a solid substrate, maximize the surface area available for heat transfer. This significant increase in surface area allows for improved heat dissipation with lower coolant flow rates, leading to energy savings. However, designing and manufacturing these complex structures can be challenging and expensive. Precision machining and microfabrication techniques are necessary, requiring specialized equipment and expertise. Furthermore, the potential for clogging within the microchannels is a crucial consideration. Careful selection of coolants and meticulous cleaning protocols are essential to ensure the long-term reliability of microchannel heat sinks. Despite these hurdles, their energy efficiency and compact size make them an attractive option for a variety of applications, especially in microelectronics and high-power devices.

Another noteworthy development is the use of advanced manufacturing techniques to optimize heat sink designs. Additive manufacturing, or 3D printing, enables the creation of highly complex and customized heat sink geometries that are impossible to achieve through traditional methods. This opens up new possibilities for optimizing heat transfer based on specific thermal requirements. Furthermore, additive manufacturing can incorporate multiple materials within a single heat sink, allowing for tailored thermal properties and enhanced performance. The ability to precisely control the material properties and internal geometry at a microscopic level offers unprecedented flexibility in heat sink design. The main current limitations reside in the speed of the 3D printing process, which can be significantly longer than traditional manufacturing methods, which might also create restrictions based on the material used in the printing, but with the continuous advances, it promises to revolutionize the way heat sinks are designed and manufactured. The high initial investment for the 3D printing technology, might be another limitation.

Emerging cooling technologies aren't just about innovative materials and designs; they also involve the integration of smart control systems. The incorporation of sensors and actuators enables real-time monitoring of temperatures and automatic adjustment of cooling parameters. This closed-loop control system ensures optimal cooling performance while minimizing energy consumption. For instance, smart algorithms can analyze temperature data and dynamically adjust fan speeds or coolant flow rates based on the real-time thermal demands of the system. Furthermore, predictive maintenance capabilities, based on data analysis and machine learning, can anticipate potential failures and schedule preventative maintenance, maximizing the system's longevity and reliability. The complex algorithms needed for the smart control systems could be demanding on the system computational power, which limits the scalability of the method. However, ongoing development in both hardware and software may mitigate these challenges.

7.1 Thermoelectric Cooling

Thermoelectric cooling offers a unique approach to temperature control, leveraging the Peltier effect—a phenomenon where heat is absorbed or released at the junctions of two dissimilar conductors when an electric current is passed through them. Unlike traditional cooling methods relying on refrigerants or moving parts, thermoelectric coolers (TECs) are solid-state devices, offering advantages such as compactness, silent operation, and precise temperature control. This makes them ideal for applications requiring precise temperature regulation in small spaces where traditional methods are impractical or undesirable.

These devices consist of multiple thermocouples, typically made of bismuth telluride alloys, connected electrically in series and thermally in parallel. The arrangement maximizes the temperature difference generated between the hot and cold sides, creating a significant cooling effect. When a direct current flows through the thermocouples, one side experiences cooling while the other side heats up. This temperature difference is directly proportional to the applied current and the efficiency of the materials. Precise control of the current enables precise control of the temperature.

While TECs offer many benefits, their efficiency is typically lower compared to traditional vapor-compression refrigeration systems. The coefficient of performance (COP), a key metric indicating the ratio of cooling power to electrical power consumed, is significantly lower in TECs. This lower efficiency results in higher energy consumption for a given cooling capacity. However, advancements in materials science are steadily improving the COP of TECs, making them more energy-efficient and competitive. The use of novel materials with higher thermoelectric figures of merit is a key research area, and ongoing investigation promises to increase performance in the near future.

The design and application of TECs demand careful consideration. The heat generated on the hot side needs effective management. Without proper heat sinking, the temperature rise on the hot side can drastically reduce the cooling capacity and damage the device. Therefore, adequate heat dissipation strategies, such as using high-surface-area heat sinks and forced convection cooling, are critical. The proper selection of heat sinks requires a careful understanding of the thermal resistance involved and heat load characteristics. Poor heat sinking can lead to premature failure of the device and greatly reduce efficiency.

Furthermore, the choice of materials significantly impacts the performance of a TEC. The thermoelectric figure of merit (ZT), a dimensionless parameter representing the efficiency of a thermoelectric material, is a crucial factor. Materials with higher ZT values offer greater cooling capacity for a given power input. Ongoing research focuses on developing novel materials, including nanostructured materials and skutterudites, to enhance the ZT values and improve TEC efficiency. Understanding the interplay between material properties, device design, and operational conditions is essential for maximizing the cooling performance of a TEC-based system.

The application of TECs spans a broad range of fields. In medical devices, they enable precise temperature control in sensitive procedures, such as laser surgery or thermal therapy. In the industrial sector, they find use in cooling sensors and electronics, offering reliable and quiet operation. Their size and efficiency make them ideal for compact electronic devices, such as laptops and smartphones, where space and noise are major concerns. Furthermore, ongoing research focuses on leveraging TECs for more sustainable applications, including waste heat recovery and energy harvesting.

Selecting the appropriate TEC for a specific application requires a detailed analysis of several factors. The required cooling capacity, temperature difference, and operating voltage must be carefully considered. The available space and the ability to efficiently dissipate heat from the hot side also play crucial roles. Manufacturers provide detailed specifications, including the maximum current, voltage, and temperature differences achievable with each device. Understanding these specifications is crucial for selecting a device that meets the requirements of the application while ensuring long-term reliability.

Despite these challenges, TECs provide a compelling alternative in various scenarios. Their inherent advantages–compactness, silent operation, and precise temperature control–make them highly suitable for niche applications where other cooling methods are less effective. Ongoing research and development efforts focused on improving material properties and system design promise further advancements, potentially expanding their role in larger-scale cooling applications in the future. The potential for improved energy efficiency, particularly as material science and microfabrication advance, makes TECs a technology worthy of continued attention and investment in research and development. Their future role in sustainable cooling solutions remains promising.

7.2 Advanced Materials

Graphene, with its exceptional thermal conductivity exceeding 5000 W/mK, surpasses copper by an order of magnitude. This remarkable property makes it an ideal candidate for heat spreading in high-power electronic devices. Imagine a microchip, its heat quickly diffused across the graphene layer, preventing localized hotspots that could lead to performance degradation or catastrophic failure. The inherent flexibility of graphene also opens up possibilities for conformal cooling solutions, adapting seamlessly to complex device geometries, unlike rigid metal heat sinks. However, the scalability and cost-effectiveness of graphene production remain challenges that need addressing for widespread adoption. Further research focuses on enhancing its integration with other materials and exploring novel fabrication techniques for mass production. This addresses concerns about long-term stability and manufacturability of graphene-based cooling solutions.

Diamond's thermal conductivity, ranging from 1500 to 2200 W/mK depending on quality and purity, also makes it a compelling material for high-heat-flux applications. While its cost is higher than graphene's, its superior hardness and chemical inertness offer advantages in certain environments. Imagine the precision of diamond machining allowing for the creation of micro-channels for increased surface area and enhanced heat transfer, making it suitable for applications where extremely precise cooling is needed. The use of diamond is particularly relevant in high-power laser systems or advanced microprocessors where the heat generation is extreme and necessitates robust, high-performance cooling solutions. Challenges related to the cost and difficulty of manufacturing large, high-quality diamond components are addressed through new chemical vapor deposition methods.

Boron nitride nanotubes (BNNTs), structurally similar to carbon nanotubes but with superior thermal stability at high temperatures, offer a different avenue for advanced cooling. Their thermal conductivity is comparable to that of carbon nanotubes, presenting an alternative with enhanced resistance to oxidation. Picture these nanotubes forming a network within a composite material, acting as highly efficient pathways for heat dissipation. Their small diameter and high aspect ratio maximize heat transfer by increasing the overall surface area, which is especially useful when considering the size and scale of modern electronic components. The challenge lies in controlled synthesis and scalable integration into functional devices. Researchers are actively exploring new techniques to disperse these nanotubes uniformly in a matrix material, optimizing their alignment for maximal heat conduction. This promises substantial improvements over current cooling methodologies.

Silicon carbide (SiC), a wide-bandgap semiconductor with excellent thermal properties, offers advantages in high-temperature environments. Its thermal conductivity, around 490 W/mK, makes it a strong contender for power electronics and high-temperature applications. SiC's ability to withstand high temperatures and voltages means it's ideal for advanced power converters and electric vehicle inverters. SiC's exceptional performance at high temperatures translates to more efficient and robust cooling solutions. The combination of SiC substrates and novel heat sink designs based on advanced materials creates an overall system with higher thermal resilience. Further development in the areas of cost reduction and manufacturing scalability will facilitate broader deployment of SiC-based cooling technologies.

Metal matrix composites, combining metals with ceramic reinforcement particles or fibers, offer tailored thermal and mechanical properties. Imagine combining the high thermal conductivity of copper or aluminum with the superior strength and stiffness of ceramic reinforcements, creating a material with optimized heat transfer and structural integrity. The result is a lighter, stronger, and more efficient cooling system. These composites can be designed for specific applications, enabling precise control over the final thermal and mechanical performance. However, optimizing the dispersion and interfacial bonding of reinforcements within the matrix material is essential for achieving the desired properties and consistent performance. This area requires continued research and development to ensure the widespread adoption of these advanced composite materials in thermal management. The future may see specialized composites designed for specific cooling applications across numerous industries.

Aerogels, exceptionally porous materials, offer intriguing possibilities in thermal insulation. Imagine a thin layer of silica aerogel acting as a highly effective barrier to heat transfer, significantly reducing the overall thermal load on the cooling system. Their extremely low thermal conductivity, even lower than that of air, arises from their high porosity and the presence of nano-scale air pockets. Aerogels can be integrated into existing cooling systems to enhance their efficiency and reduce energy consumption. However, their fragility and cost remain obstacles to overcome for widespread application. This necessitates research into developing more robust and scalable production methods of aerogels, and explore novel compositions, such as carbon-based aerogels, to improve mechanical properties while maintaining excellent thermal insulating characteristics.

www.ingramcontent.com/pod-product-compliance
Ingram Content Group UK Ltd.
Pitfield, Milton Keynes, MK11 3LW, UK
UKHW020156290926
15471UKWH00005B/911

9 798349 286292